AF682344

THE EXTRACTION OF FREE GOLD

Al McGowen

THE EXTRACTION OF FREE GOLD

Al McGowan

Copyright, 1973, Al McGowen

Photographs by Jim McGowen and H. Glenn Carson

Published by H. Glen Carson Enterprises, Ltd.,
801 Juniper Avenue, Boulder, Colorado 80302

TABLE OF CONTENTS

A FORWARD FROM THE PUBLISHER

The extraction of free gold, or at least the attempt at its extraction, is not new. Men down through the centuries have tried to take the bits and pieces of precious, shining metal from their hiding places. The search for gold, indeed, has often proven to be a more difficult task than even searching after the proverbial needle in a haystack. Yard after yard of placer material may hide only a few tiny flakes of gold, minute specks in a gigantic mass of surrounding material.

Man discovered gold and developed a love for it many long years ago. It is the first metal mentioned in the Bible. Early man probably picked up shining nuggets along streambeds, admired them, and kept them even before they discovered and used copper. The golden fleece of Jason is not merely legend. Ancient gold-seekers in what is now Turkey found that a sheep's fleece made an excellent gold entrapment when placed under water in a goldbearing stream. The fibrous wool trapped particles of gold. Indoor-outdoor carpeting just works a lot better, but the ancients had no way of knowing that fact.

What Al McGowen has to say about the extraction of free gold is simply a straightforward presentation of facts learned over a lifetime of mining and searching out ever-better methods of taking the shining metal from its surrounding sands and gravels. There is no magic to it, although it almost seems so at first glance. What Al has done has simplified the complicated, made practical the wild daydream, and made economical sense out of financial folly. This book has long needed compiling, and it is with a great deal of pride that I have a small share in presenting these methods to those interested in the taking of gold.

H. Glenn Carson

EXTRACTING FREE GOLD

This part of the book deals with taking free gold, as a bi-product, from sand and gravel mining operations in the front range area of Colorado. The following information is based on the actual recovery of gold at sand and gravel plants located on the St. Vrain River at Longmont, on Boulder Creek east of Boulder, Clear Creek east of Golden, and on the South Platte River north of Denver. All of these rivers drain areas which actively supported gold mining at one time. Similar areas should offer comparable results, but obviously , many drainage systems will not contain enough gold to make attempted recovery profitable. Sufficient gold has to be present if there is to be profitable recovery. Extensive sampling of the sand and gravel site must be carried out to determine the economic feasibility of saving gold as a bi-product.

The size of the gold particles recovered at these sites varies from a very fine 400 mesh (a screen which has 400 openings to the inch) to 10 mesh (10 openings to the inch). The gold passing through 400 mesh is extremely fine, whereas the leafy flakes barely able to pass through 10 mesh are quite visible. Occasionally larger particles are recovered.

The gravel plants, as a rule, turn out different sized products. These are 1¼ inch aggregate, ¾ inch aggregate, and the finer sand, which passes through a ⅜ inch screen. Since the free gold is fine, all of the gold is in the sand product. A most important rule of milling and heavy metal separation holds true here: **There can be no concentration without size classification.** The "rag plant" used to catch the gold is set up on this ⅜ inch screened (classified) product.

A rag plant is a steel launder or sluice box. It is a large, flat-bottomed trough which is set at enough of an angle so that water can carry sand from the upper to the lower end. This launder is used to transport the sand, but it is also the place where any gold can be separated from the sand. Rag plant sizes vary according to how much sand is processed per hour, and how much water is used to carry the sand. A large operation (200 tons per hour) may require a rag plant eight feet wide, fifty feet long, and ten inches deep. Sand and gravel plants handling smaller tonnage do not require as large a rag plant. Ideally, however, to ensure maximum gold recovery, the rag plant should be at least thirty feet long. Shorter distances are very likely to lose varying amounts of gold. The bottom of the rag plant should be flat and smooth.

Sand and gravel plants make their money by producing sand and gravel. Anyone setting up a rag plant must always keep that fact uppermost in their mind. Everything on the rag plant must be installed so that it cannot interfere in any way with the sand gravel operation or cause a bottleneck at any time. For this reason the water flow and pitch of the rag plant must be tailored to sand and gravel production rather than gold recovery.

A highly efficient method worked out over years of actual operation is to use indoor-outdoor carpet on the bottom of the rag plant, with steel punchplate riffling directly on top of the carpet.

The carpet catches the gold particles as they are carried down the rag plant. Indoor-outdoor carpet has definite advantages over burlap, wool blankets, and wool rugs, which have been and often still are used. Being a plastic, the carpet does not rot or fall to pieces as do the other materials. The "nap", or surface fibers are very resilient. They are not permanently crushed by the weight of the sand and water passing over them. Being plastic, these fibers do not "wilt" when they become wet. Although the initial cost of this carpet is high its efficiency and long life make its use more economical than other materials.

Punchplate is steel plate with holes punched out of it. It serves two purposes. First, it holds the carpet flat against the bottom of the rag plant. Secondly, the holes in the punchplate cause a riffling action, or a turbulence, in the water and in the sand being carried by the water. It is this turbulence that allows the gold to be dropped from the flow of sand and picked up by the fibers of the carpet. A punchplate with square holes seems to make the most effective riffle. Round holes work, but not as well. Punchplates which are one-quarter inch thick, with one and one-quarter inch square holes half an inch apart have worked very successfully. The rows of holes should be staggered. One hole should not be directly in line above another hole in the direction the sand and water are flowing. In this way all material has to pass over fiber-filled holes. These punchplates should not be too big to be handled by one man.

It is important that the plates be pulled down tightly on the carpet and the bottom of the rag plant. The tightness forces the fibers of the carpet to stand high up in the holes. To make this required tightness, half-inch bolts are welded to the bottom of the rag plant along where the edges of the punchplate go. These bolts should be spaced about two feet apart, and form a line of bolts up and down, in the direction of the flow along the rag plant. Holes are drilled in two-by-four inch lumber so that the lumber can be placed over the rows of bolts. When dropped into place the two-by-fours cover the edge of the punchplate. Washers and nuts are then put on the bolts and the lumber is pulled down until it forces the plate into the carpet. Because of the width of most rag plants and the weight of the punchplate, two or three rows of punchplate will be used across the width of the rag plant. A single row of bolts and one two-by-four between two rows of punchplates can be used to hold both rows of plates securely down.

Separating gold from the sand on the rag plant is based on a very simple principle. GRAVITY. The heavier particles tend to be on the bottom of the material flowing down the rag plant, and they tend to move more slowly than lighter particles.

Most of the "sand" moving down this trough is quartz, feldspar, mica, and other relatively light materials. A small percent, usually less than two percent, is magnetic iron, or "black sand". An extremely small percent is gold. It is the difference in the specific gravity of the gold and the other materials that makes separation possible. Specific gravity is simply the weight of a substance in relation to the weight of the same volume of water. Gold has a specific gravity of 19.3, therefore a cubic inch of gold is 19.3 times as heavy as a cubic inch of water. The lighter quartz fraction has a specific gravity of less than three. The blacksand fraction has a gravity less than six. Gold, with a gravity of nineteen point three is more than six times heavier than the light quartz fraction and more than three times heavier than the black sand fraction.

This means that as the material moves down the rag plant the gold will tend to settle into the carpet fibers sooner and easier than the black sand and quartz. The turbulence caused by the riffling increases the chance that the gold particles will settle and not be swept with the sand off the rag plant.

In the many miles from their origin these particles of gold have moved, and over the millions of years it has taken them to move, these particles have become flattened and irregular. These irregular particles are held by the nap of the rug. As the water and sand flow over them the particles even work their way down between the fibers of the rug. As the gold collects in the square holes of the punchplate, the depth of the hole serves to protect the gold particles from being carried away.

Several factors can prevent the rag plant from working properly. First is the pitch, or angle of the rag plant. The pitch is determined by the amount of fine sand in the deposit being worked. The more fine sand there is the steeper the rag plant must be. The pitch is usually between one and one half inch and two inches per foot. If the rag plant is not steep enough it will "sand up", or collect on the bottom of the rag plant. This situation interferes with the sand production of the plant. The pitch would have to be changed or more water would have to be added to the sand flow to ensure its movement.

A situation more apt to hinder a gold recovery operation, even though it would not hinder the sand production, is when the black sand is "sanding up" on the rag plant bottom. The lighter sand moves properly, but the pitch is not steep enough to carry the heavier black sand. In this case the holes in the punchplate soon fill with black sand, leaving no place for the gold to collect. The gold is then swept off the rag plant. To correct this situation the pitch of the rag plant must be increased, made steeper, or more water added to the sand.

Conversely, if the pitch of therag plant is too steep, the gold may not be able to settle into the carpet, or be able to remain on the carpet once it has reached it. This situation may sometimes be corrected by using less water. If not, the pitch must be decreased slightly.

It can easily be seen that water affects the operation of the rag plant much the same as does the pitch. If it is possible to regulate one or both of these factors, most problems in gold separation on the rag plant can be solved. If regulation of only one factor is possible solutions to problems are still possible, but often more difficult.

If the bottom of the rag plant is uneven the plates may not pull down tightly on the carpets. This is undesirable, for the gold that settles will not be protected in spots and may move down the

Here is the ragplant after the carpets have been changed. Note the space which was left between the punch-plates to make them easier to pickup. This ragplant is approximately ninety five inches wide. The section to the left has been added on to make the ragplant wider.

To change the carpets, the two-by-fours are removed and the punchplate riffling are taken up. The carpets are then rolled up and clean carpet laid. After the plates and timber are replaced, the ragplant is ready for another week's operation.

rag plant and be lost.

Because all of the particles moving down the rag plant are not the same size, some of the larger iron and quartz particles may settle and become trapped in the carpet and riffling. Likewise, some of the very small gold particles will not be caught, and will be lost.

The carpets need to be changed only once a week on most rag plants. After a week's time there should be spots of carpet showing in the square holes of the punchplate. The carpet should not be left on the rag plant long enough for the holes to be filled with gold, for some gold would then be lost. Heavier accumulation should always be found at the upper end. Gold at the upper end should be coarser. The amount of gold should diminish and become finer as one moves down the rag plant. If there is as much gold on the lower half of the rag plant as there is on the upper half, the rag plant is not operating properly. This situation usually occurs if the rag plant is not long enough. If possible the rag plant should then be lengthened to ensure all of the gold a chance to collect on the bottom and not be washed off the rag plant.

To clean the carpets the two-by-fours are removed and the plates taken up. Each punchplate should be carefully brushed off over the carpet, to ensure that any gold sticking to it is collected. Next, the carpets are rolled up and put into containers. Care must be taken at this time that the material on the carpet is not knocked off and lost. Any material which has accumulated under the rugs is swept up and put into the containers. Clean rugs are put down, the punchplates replaced and pulled down tight with the two-by-fours. The rag plant is then ready for another week's operation.

Drying the rag plant rugs is the first job at the cleanup mill. Because indoor-outdoor carpet is plastic, the fibers do not absorb moisture. Only the material on and in the rug has to be dried. The carpets are hung from heavy clamps over a large catch tank. As the carpets dry some of the material sticking to them will fall into this tank. Circulating air and heat can be used to speed this drying process. When the rugs are totally dry they are put into a rotating tumbler along with heavy pieces of rubber. The rubber pieces fall against the rugs as they rotate, helping to knock the gold, sand, and iron from the rugs.

A tumbler three feet in diameter and three feet long will accommodate three rugs, each thirty inches wide and eight feet

These wet carpets off the ragplant will soon be hung up and dried.

The rugs are hung from clamps and warm air is circulated around them to quickly dry them. Notice the pattern left by the punchplate. The dark material is magnetic iron.

This is the tumbler which is used to remove the dried material from the indoor-outdoor carpet. It is made from an old commercial laundry drier.

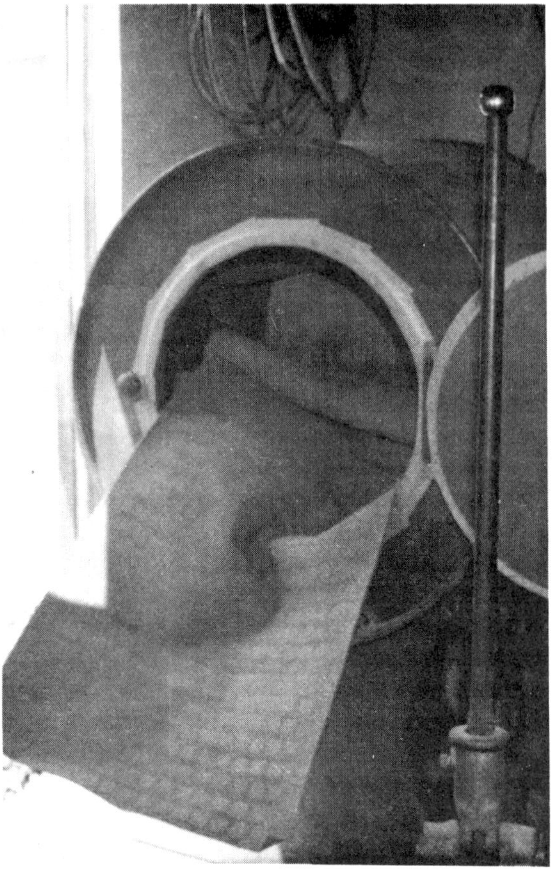

A rug is removed from the tumbler after being cleaned. The dark pattern on the carpet is from the punchplate which held the carpet down on the ragplant.

long. The rotating drum of the tumbler should be a stainless steel punchplate with quarter inch holes in it. The material passes through these holes after coming out of the carpet. A hopper to collect the material can be built under the tumbler. The tumbler should be tightly made so that no dust leaks from it when the rugs are being tumbled. If not, the area will be so dust-filled it will resemble a Mojave Desert dust storm.

The next operation in the gold clean-up is to put all the material recovered from the rugs across a concentrating table. Materials are separated according to their specific gravities on the concentrating table. To the layman a concentrating table could be said to be a controlled mechanical panning machine. Of course it is much more rapid and efficient than a man panning gold. On the concentrating table one has control of the amount of water being used, the amount of material being put onto the table, the size of the material, the speed of the table's vibration, and the absolute control of the pitch or angle of the table.

The table is basically a flat, rectangular surface with raised pieces or riffles running the length of it. The table is pitched slightly in the long direction and more steeply pitched in the other direction. The steeperpitch across the table is adjustable. The size and gravity of the material to be put across the table determines the pitch at which the table is set. The heavier the material, the steeper the pitch must be. The feed, or material to be put across the table, drops onto the table at the highest corner. Water is added with the feed, and all the way across the table. The table's motion comes from an eccentric "head motion", which moves the table back and forth in the long direction of the table. A small table is moved about three-eighths of an inch by the head motion, many times each minute.

As the feed drops onto the wet table top, the lighter materials move quickly down the steep pitch of the table. The heavier materials, such as gold and iron, hang higher on the table. The motion of the table moves the material in the long direction of the table. A few inches from the top of the table the first riffles start. The first ones are short, each consecutive one longer than the last, until three-quarters of the way down the table the riffles extend the length of the table. Each riffle is tapered from a quarter of an inch at the high end of the table to a sixteenth of an inch at the low end of the table. The material moving down the table is stopped by these riffles and moves along them, the lighter material backing up and crossing the riffles as it becomes thinner, then picked up by the next riffle. By the time the material has reached the far end of the table it is in a line running roughly from the high corner of the table towards the low corner of the table.

The bulk of the concentrating is done on the concentrating table. This table was a Wilfley head motion and a stainless steel deck with brass riffles. The feed box is in the upper right hand corner of the table. The water box runs across the top of the table. Two "highgrade riffles" can be seen extending the length of the table. A streak of black iron can be seen running roughly diagonally down the table. Below this streak a wider streak of sand can be seen. Most of the gold will be carried to the end of this table by the highgrade riffles.

Looking at this line of material, one will see the upper part of it contains the heaviest material on the table and the lower part the lightest material on the table. By carefully splitting this line of materials as it falls off the far end of the table, a separation of materials by their specific gravities can be made. The concentrating table makes a good separation, but there will be some contaminates in each fraction split off the table.

Before the material from the tumbler can be put on the concentrating table it must be screened. As mentioned earlier, the maximum size of the gold particles in these Colorado deposits is approximately 10 mesh. A 10 mesh screen, therefore, is used to screen the material from the tumbler. A vibrating feeder is used to feed the screen. A rheostat on the feeder controls the rate at which the material is fed onto the screen. After screening, the 10 mesh material moves onto the concentrating table. The vibrating feeder is set to feed exactly the right amount to the table, for too much or too little material will cause the table to work improperly.

If one examines the screened material from the tumbler as it comes off the end of the concentrating table it will be seen that the quartz and other lighter materials are dropping off at the bottom of the table. In the upper part of this iron streak we will see some flakes of gold mixed in with the iron, and the top of the iron streak we will see a thin streak of gold. The iron streak is split below the area where the gold particles are mixed with the iron to ensure that all the gold is saved. Material below this line is discarded, and is called the "tail". Material above this point is saved and called the "concentrate", or "con".

Thus, with the help of the table, practically all of the quartz, feldspar, and so forth, and most of the black iron, has been eliminated. The concentrate should contain all of the gold collected on the rag plant, and only a small percentage of the iron collected. If the table concentrate is examined closely, small amounts of substances other than gold and iron will be found. These may include zircon, lead, and other heavy substances.

The next step is to dry all of the table concentrates in preparation for the magnetic separator.

The principal contaminate in the concentrate is now black or magnetic iron. The magnetic separator, as the name implies, is used to separate magnetic substances from non-magnetic substances. Gold is not magnetic. Black iron is magnetic. Most of

The A.C. magnetic separator pulls the magnetic iron from the center of the belt to both edges of the belt. The material which is left in the center of the belt is nonmagnetic. As the material falls off the end of the belt it is collected in separate containers. The two outside containers will contain only magnetic iron while the center container will contain gold and other nonmagnetic materials which may be present.

The three phase A.C. magnetic separator is used to separate magnetic iron from gold. To the right is the hopper and vibrating feeder. Three A.C. magnets are located beneath the white conveyor belt, one in front of the D.C. magnet and two behind the D.C. magnet. The D.C. magnet, which hangs over the belt, imparts a magnetic charge to the less magnetic iron as it passes under it. This "charged" iron is then removed as it passes over the last two A.C. magnets.

the magnetic iron will be separated out of the concentrate on the magnetic separator.

The AC magnetic separator is a fairly complicated piece of equipment. It uses 3 phase, 220 volts, in three magnets located under a conveyor belt which carries the material to be separated. The rotating field of the AC magnet sets up a magnetic field which causes the particles of iron to move. The poles of the magnets are situated under the belt in such a manner that the iron particles move from the center of the conveyor belt outwards to both edges of the belt. Eddy currents are set up in each particle of iron, which cause that particle to spin as it moves towards the edge of the belt. These spinning particles actually move in a progression of tiny jumps, or hops. Gold particles cannot be dragged to the edge of the belt by these twisting iron particles, for they are airborn most of the time they are moving.

The table concentrate, after being thoroughly dried, is fed onto the center of the slowly moving conveyor belt of the magnetic separator. Again, a controlled vibrating feeder is used to ensure that just the right amount of material is fed onto the belt. As soon as the material is on the belt it moves over the first AC magnet. Magnetic iron begins to move to the two edges of the belt. By the time the material on the belt has reached the end of this first AC magnet most of the finer black iron has been moved to the edges of the belt. Gold and other non-magnetic materials, if any are present, plus particles of iron too large to be moved the edges of the belt, or iron that was not magnetic enough to move, are left in the center of the belt.

The belt now moves under a DC magnet. This electromagnet is like the permanent magnets we all know and have played with. Any iron remaining in the center of the belt is picked up by the DC magnet. Most of the gold and other non-magnetic materials on the belt pass under the DC magnet and continue on. Iron builds up on the DC magnet until so much is clinging to it that it can hold no more. The excess iron the magnet cannot hold continues on the belt. Beyond the DC magnet are two more AC magnets. They are identical to the first AC magnet and move much of the iron that has not already been moved to the edges of the belt by the first AC magnet. Some of the less magnetic iron becomes more magnetic while passing under the DC magnet. This iron also moves to the edges of the belt while passing over the second and third AC magnets.

Three containers are placed at the discharge end of the belt. The one in the center catches the gold and other nonmagnetic material, and the small amount of iron that was not removed by the magnets. The material in this middle container is the magnetic separator concentrate. The two outside containers catch the magnetic iron which has moved to the edges of the conveyor belt. The magnetic iron in these two containers is the "tail" of the magnetic separation and is discarded.

Occasionally during this process the power to the DC magnet is shut off. The material clinging to it is dropped off onto the belt and collected in a separate container as it falls off the end of the belt. A small amount of gold has been trapped in this iron as it collected on the DC magnet. This is a second, less rich concentrate, and will be saved.

The magnetic separator used in this cleanup mill is approximately twelve feet long and thirty inches wide. The belt is ten feet long and eighteen inches wide. The three AC magnets measure 18 by 20 inches. The DC magnet covers the center four inches of the conveyor belt.

Practically all of the magnetic iron has been eliminated by this time: most of it on the separating table, the remaining on the magnetic separator. The primary concentrate consists mainly of gold, with small amounts of other non-magnetic materials. This concentrate must be carefully washed by hand to separate, by gravity, the gold from the other materials present. The final product will be clean free gold, or placer gold. The secondary concentrate contains mostly iron, with only a small amount of gold. This concentrate cannot be easily cleaned by washing, but instead must be amalgamated, then melted, and poured into a mold to form a gold "button". Both of these processes will be discussed in detail.

Gold in its natural state (free gold, placer gold, gold dust, gold nuggets, and so forth) is not now regulated in any manner by the United States government. Any private citizen of the United States may own as much free gold as he can afford or wishes to own. He may buy it from whom he wishes and sell it to whom he wishes. There are, at the present, no restrictions on such activities with free gold. One may not melt or treat it in any manner, however, that would change its form, for once gold is melted or changed from its natural state, it becomes strictly controlled by the rules and regulations of the Treasury Department of the United States government. The melting of gold and the buying and selling of melted gold are closely regulated, and most persons engaged in such activities must be licensed by the Treasury Department. Restrictions are set on how much melted gold may be held at one time, and how and to whom it must be disposed of is strictly outlined. Melted gold cannot be held as a commodity, but must be consumed or used. One cannot store it away in a safety deposit box, but can make gold jewelry, dental hardware, or plate electrical terminals with it.

Since free gold is not controlled by these stringent governmental regulations it is very much worthwhile to take extra care in the cleanup mill to produce as much free gold as possible. Besides the convenience of selling to whomsoever one wishes, placer gold can usually be sold at the world market price. Melted gold, on the other hand, must usually have a "fineness" run on it, to check its purity, and the seller is paid only for the actual weight of the gold, and not for the weight of the impurities within the gold. Gold with a fineness of 1000 is pure gold. If the fineness is 900, the material is 90 gold, and 10 impurities. The impurity usually consists mainly of silver.

The next step in the cleanup mill is the final washing of the gold. This step is done by hand. The material to be washed is the first, or richer, magnetic separator concentrate. It is composed mainly of gold, but has in it some non-magnetic impurities.

This gold concentrate is first classified or screened to different sizes. Screens of 20 mesh, 25 mesh, 30 mesh, 35 mesh, 40 mesh, 50 mesh, 60 mesh, 70 mesh, 80 mesh, and 100 mesh are used. Gold particles larger than 20 mesh are difficult to wash and are added to the concentrate to be amalgamated. Each of the different screened sizes is placed in a separate container and washed separately.

Four pieces of equipment are used in this washing process. A rectangular stainless steel washing pan, about a foot square, with sides two inches high on three sides of it. A flexible hose with a small nozzle capable of spraying a fine jet of water. A large pan or tub. A small spatula. Not a remarkable list of equipment in this supposedly technological era, perhaps, but a most practical one.

The washing pan is set on an incline of about three inches

drop to the foot, with the open end of the pan in the downhill direction. The tub is set under the pan so that all material dropping off the open end will fall into it. The washing pan may be placed in a launder or trough about two or three feet long, with the tub placed under the far end of the launder to catch material as it falls off the end of it. If set up this way, the washing pan is steadier, and the pitch is more easily set.

The water is turned on and the nozzle adjusted to shoot a small jet. A spoonful of gold concentrate is placed near the top of the washing pan. Using the water jet, this material is carefully moved towards one side of the washing pan. As the water moves this material the lighter impurities are carried down the washing pan and into the tub. Before the material reaches the side of the washing pan, the direction of the jet of water is changed to push the material towards the other side. The material is worked back and forth across the pan in this manner as long as necessary. As the material moves, the lighter impurities are carried down the washing pan into the tub below. Obviously, some of the lighter particles of gold will also be carried away by the jet of water, but they will be caught and saved in the tub. This lighter gold will later be amalgamated.

When no more impurities can be seen in the gold as it moves across the pan, the gold is washed up into a top corner and removed with a small spatula. Another spoonful of gold concentrate is placed in the washing pan and the process is repeated. A certain amount of technique is required in this washing process. One may very likely find himself re-washing an entire batch of concentrate several times before gaining an adequate technique.

Other than drying, this is the last operation to be performed on the gold. All the remaining impurities must be removed at this time. After drying, the result is the finished product, placer gold.

Amalgamation is a process of alloying certain metals with mercury. The process has been known and used for centuries. Mercury, or quick-silver, is a heavy metal which is liquid at room temperature. Gold, silver, copper, and zinc are easily alloyed with mercury, and the alloys which are formed are called amalgams. Other metals, such as iron, do not alloy with mercury. The process of amalgamation, therefore, can be used to separate gold, which is easily alloyed with mercury, from iron and the other impurities in the concentrate which will not alloy with mercury. The mercury is used to pick up the particles of gold in the concentrate and bring them together. A heavy gold amalgam is formed which can be separated from the impurities by gravity.

If particles of gold are to be alloyed with mercury, they must be clean and free of any surface scale or rust build-up. These clean gold particles must then be thoroughly mixed with the quicksilver. To ensure this, an apparatus called an amalgamator is used. It consists of a cast steel bowl, about ten inches in diameter and six inches deep; a cast steel shoe, or mauler fits into this bowl. The shoe rotates and effectively grinds and cleans up the gold when it is added to the bowl. When mercury is added, the shoe thoroughly mixes the gold and mercury. Amalgamators such as this have been used for a hundred years, and cruder versions for several thousand years, and are still the most effective type to use. This is so because such devices ensure a one hundred percent recovery of all gold put into them, and because they will not "flour" the quicksilver.

"Flouring" the quicksilver simply means that the liquid mercury becomes separated into many small drops which become dispersed throughout the concentrate and are most difficult to get back together, especially after they become "loaded". Loaded quicksilver is that which has picked up gold. As the gold is picked up, the quicksilver loses its fluid nature and becomes stiff. If the quicksilver is floured, or spread through the concentrate, and loaded, it is difficult to collect all of the small amalgam particles and bring them together. Unless these particles are brought together, no effective separation of the gold amalgam from the rest of the material in the concentrate can occur.

These amalgamators are scaled down copies of the old Spanish arastras first used in Mexico in the 1500's to grind ore.

A gearhead motor is used to turn the heavy shoe of the amalgamator at thirty r.p.m. The shoe is not held rigid from above, but is allowed to rest on the bottom of the bowl. A splined shaft rotates the shoe but is not fastened to it. This ensures that the shoe moves evenly on the surface of the bowl and that the weight of the heavy shoe is constantly pressing down on the bowl. This is important, because the amalgamation of the gold will take place between the bowl and the moving shoe. The shaft

This small amalgamator has been mechanized. The plined shaft which turns the shoe can be raised and lowered. The shoe turns about 30 rpm.

Here the shoe has been removed from the amalgamator. Amalgamation takes place between the shoe and the bowl as the gold is ground and mixed with the mercury.

turning the shoe can be raised and the shoe removed from the bowl, when necessary.

The material to be amalgamated is added to the bowl while the shoe is turning. Water is added until the bowl is about half full. To ensure the gold being clean and ready to alloy with mercury, this material is ground for twenty minutes before the mercury is added.

To determine how much mercury is needed one must estimate how much gold is in the bowl. Use twice that amount of mercury. If one is in doubt as to how much gold is present, an excess of mercury should be used. It is important to have enough mercury present so that after the amalgamation has taken place, the gold amalgam and the excess quicksilver form a puddle of liquid which can be easily poured. If insufficient mercury is used, the gold amalgam cannot ride in the liquid and will become dispersed throughout the impurities in the concentrate. It will not be easily separated from them. After the gold has been cleaned about twenty minutes the mercury is poured into the amalgamator. The material in the bowl should never come up past the middle of the shoe, for any material above that would not be completely amalgamated. In most cases amalgamation will be completed in forty-five minutes.

The motor is then shut off and the shaft that turns the shoe is raised. The shoe is taken out of the bowl and set in a large pan. A gold pan works nicely. In cleaning the amalgamator and separating the mercury and gold amalgam from the rest of the material, extreme care should be taken. Liquid mercury can easily be spilled or lost without one even knowing it. Always work over a pan or tub filled with water. In this way, in case of an accidental spill, the material will not be lost. Anyone who has ever played with mercury knows that once it is spilled on the floor, there is no way it can all be collected again. The shoe is placed on its side in the pan and carefully washed off. The bowl is next emptied into the pan, then laid on its side and any remaining material is carefully washed into the pan.

All of the material from the amalgamator is placed in a gold pan and the lighter "insul", or iron and other impurities, are panned off by hand. One must be very careful while panning off the lighter materials not to spill any mercury. Again, the panning should be done into another pan to ensure that if spilled, the mercury can be recovered.

When enough insul has been removed so that the mercury is visible in the bottom of the pan, the mercury is carefully poured into a two-cup size pyrex measuring cup. These are preferable to glass beakers. They are heavier and not as easily broken. They have handles and pouring spouts, are not expensive, and can be

A chamois skin is used to separate gold amalgam from mercury. Here the mercury and amalgam have been placed in a chamois and the edges of the skin twisted together.

purchased at any hardware store. Check the pan to make sure that all of the mercury is in the measuring cup. If some mercury remains in the pan, enough insul should be panned off so that the remaining mercury can be poured into the measuring cup.

The measuring cup at this time contains mercury, the gold amalgam being carried by the mercury, and a small amount of insul that was carried into the cup when the mercury was poured in from the pan. To remove this small amount of insul, fill the measuring cup with water. Pour the water and mercury into a second measuring cup.

Some of the insul has remained on the bottom and the sides of the first cup. Wash this cup out carefully, pour clean water over the mercury in the second cup, and pour the contents of the second cup back into the first cup. The second cup is washed out. The procedure is repeated until all the insul has been removed. The cleaned product now contains only mercury and the gold amalgam being carried by the mercury.

The next step is to separate the gold amalgam from the excess mercury. This is done by squeezing the mercury through a chamois skin. The chamois skin must be a prime, thin skin, not the heavier type used at filling stations for cleaning windows. Mercury cannot pass through the pores of the thicker skins. The thinner skins can usually be purchased at a drugstore.

The chamois is placed in a container of water and allowed to soak for several minutes. Fill a four-cup pyrex measuring cup three-quarters full of water. When soaked, remove the chamois skin from the water, spread it out, and place it over the top of an empty two-cup size measuring cup. Push the skin down into the empty cup to form a pocket to hold the mercury and gold amalgam. Pour the mercury and gold amalgam into this pocket. Then the sides, which are draped over the edge of the cup, must be carefully gathered up and pulled together. The chamois should now resemble a sock with the heavy mercury and gold hanging in the toe of it.

Now place the chamois in the water in the larger measuring cup, being careful to hold the gathered edges tightly. Water is used in the cup so that if a leak should occur in the skin, the mercury will not shoot out, be spilled, and lost. Holding the gathered edges of the skin tightly with one hand, begin twisting the skin as you can. Continue twisting down toward the mercury and amalgam. As the skin tightens around the contents, mer-

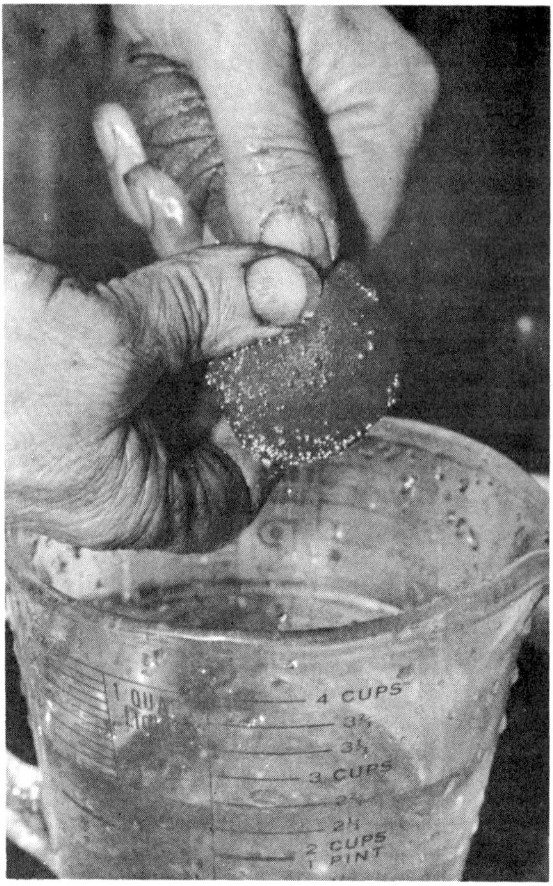

By twisting the chamois liquid mercury can be forced out through the pores of the skin. The particles of gold amalgam are too large to pass through the skin. After all the unalloyed mercury has been squeezed through the skin, the skin is opened and the amalgam removed.

cury will be forced through the small pores of the skin and fall to the bottom of the cup. The particles of gold amalgam are too large to pass through the pores, and remain in the chamois. Continue squeezing until no more mercury can be squeezed through the skin.

Now remove the skin from the water. Carefully squeeze any excess water from the skin. You will now feel a hard lump in the bottom of the skin. This is the gold amalgam.

Place the chamois on a flat surface and carefully open up the skin. When completely opened, turn the chamois over and allow the amalgam to fall onto a clean piece of paper. Using your fingers, work all the amalgam into a ball. Place this ball in a container with a tight lid. Wash your hands as soon as you finish doing this. Great caution should always be taken when working with mercury, and the substance should not be touched more than necessary. It should **never** be placed near the mouth. When storing it, keep it in an unbreakable container with a tight-fitting lid, or with a layer of water covering it. Mercury vaporizes at room temperatures if not thus contained, and is extremely toxic.

The ball in the container is now an amalgam button of roughly fifty percent gold and fifty percent mercury.

The next step is to separate the mercury from the gold in this approximately fifty-fifty button. This is done by distilling off the mercury. Mercury boils, or turns into a gas or vapor at about 675 degrees Fahrenheit. Gold, on the other hand, remains a solid until heated to about 1900 degrees Fahrenheit. The amalgam button, therefore, can be heated until the mercury is driven off without melting the gold that is present.

This is done in a specially built mercury retort. This is one step that should never be attempted by the inexperienced, or done with questionable equipment. Mercury vapors can kill very quickly. Many an old miner died in his cabin while trying to distill his amalgam buttons atop his cook stove.

The mercury retort is composed of a steel pot and a steel lid which is machined to fit tightly. Never attempt to distill mercury in a glass retort. A metal pipe extends out of the retort lid, bends, and extends down at an angle. This pipe is water jacketed, and at the end of the pipe extends into a container of water. Cold water is run through the water jacket, entering at

The small shiny ball of material is an amalgam button. It is composed to mercury and gold which are alloyed together. The next step is to heat the amalgam button in a mercury retort to separate the gold and mercury. The flask in the back is used to store mercury in. Always store mercury in a sturdy container with a tight fitting lid.

the lower end and discharging at the upper end. The pipe, with the water cooling it, serves as a cooling chamber. It is within this pipe that the hot mercury vapor condenses, or is cooled back into liquid mercury. The retort is set up under a hood, which has an exhaust fan to carry away any stray fumes should a leak occur.

Again, take no risks while using mercury. It is dangerous stuff.

The amalgam button is placed in the retort pot. Tighten down the lid, and start water through the water jacket. Heat the retort. When the temperature inside the pot reaches approximately 675 degrees Fahrenheit, the mercury begins to vaporize. This vapor moves into the pipe, and there at once begins to cool. By the time it reaches the lower end of the pipe it has cooled sufficiently to become a liquid again, and runs out into the container of water.

To ensure that all the mercury is driven from the button, it is wise to heat the retort pot up to about 1100 degrees Fahrenheit.

Turn off the heat after no more mercury can be seen dripping from the end of the pipe. Allow the pot to cool. After the retort pot has cooled, open it, and remove the gold. The gold will be stuck together and have irregular holes and pock marks over it. Because of these features it is called a "sponge button".

The distilled mercury should be stored in a safe container, of course, to be used again another time.

The final step is to melt the gold in the sponge button and pour it into a mold. A furnace that will reach at least 2000 degrees Fahrenheit is used to melt the gold. The furnace should have a hood over it, so that if a small amount of mercury remains in the sponge button, the vapors will be safely exhausted by the fan in the hood.

The sponge button is placed in a crucible and covered with a couple of teaspoons of flux. This flux is borax glass or a mixture of borax glass and soda ash. Flux serves as a cover over the gold as it melts, and makes the melting easier and faster.

Preferably, the crucible should be graphite. A fire clay crucible can be used, but the graphite crucible is far less likely to break and can be used many times. The fire clay crucible is consumed by the molten metal and may crack or break. It should not be used more than once or twice.

The crucible is placed in the furnace, and left there until it is

The mercury retort is used to separate mercury from gold in the amalgam button. The button is placed in the pot at the left. The burner is fired and when the termperature inside the pot reaches approximately 675 degrees F the mercury begins to vaporize. The vapor is cooled as it moves through the water jacketed pipe and drips into the beaker at the end of the pipe as liquid mercury.

certain that all of the gold in it has melted. The crucible then is removed from the furnace with long tongs and poured into a gold mold. These molds are made of cast iron and have a number of conical shaped holes in them. When the contents of the crucible are poured into one of these holes, the heavier gold goes to the bottom, the lighter flux, which is also melted now, covers the top of the gold.

After the gold has solidified the mold is turned over and the gold and the flux drop out of the hole. When cooled sufficiently the flux is broken away from the gold. If any flux remains stuck to the gold it can be easily removed in hot, soapy water.

The final product is a cone-shaped gold button.

After the mercury has been retorted from the amalgam button, the gold is melted and poured into a mold. The mold in the background has cone shaped holes and is made of cast iron.

SAMPLING A PLACER

Sampling is an important first part of any mining program, whether large or small. A good sampling program, even though it can run into some cost, will save money in the long run. No mining venture should be started on old stories, a quick look at the property, or hope. Instead, a careful exploration program should be followed. A most important part of this exploration is the sampling of the deposit.

The main objective of sampling is to decide if the deposit can be worked for a profit, but other important things can also be learned from a good sampling program. Knowing these things may change a borderline deposit into an economically profitable deposit.

By nature, most placer deposits are irregular and inconsistent. Because most deposits were formed by stream action depositing material, they can be and usually are as irregular as the flow and form of the river which deposited them. Certain parts may have relatively little gold, while other spots in the deposit may contain much gold. By carefully sampling, one may be able to pinpoint where the better parts of the deposit are most likely to be found.

Layering of the deposited material, or stratification, is sometimes very apparent in placer deposits. Careful examination of these layers may show that some contain much more gold than the others. A base layer may be found, below which little or no gold will be found. Knowing this could save money, for then the lower part of the deposit would not be mined. Likewise, the upper layers may not contain much, or no gold, and might be stripped off and only the lower strata mined.

It is important to know the range of the sizes of the gold particles when working a placer deposit. Are there enough large nuggets to make hand picking profitable, or only an occasional nugget? Will the riffles in your sluice box catch all the gold, or will the smaller particles be carried off, requiring the tail to be rescreened to a smaller size and put over a second sluice with less water? There may not be enough gold in a certain size range to make it profitable to recover that size range. A good sampling program may answer many of these questions and give you information which will enable you to plan just what type of recovery will be economically profitable. Many placers have been worked to recover coarse gold, with no attempt to save the finer gold. And in many cases the finer gold, if it had been saved, would have amounted to more than the coarse gold did. A good sampling program then, should give you an idea of how much gold is present, but also an idea of how the deposit should be worked if it shows economic worth.

A good sample is a representative sample. Because of the inconsistency of most placer deposits, each sample should be representative of the area from where it was taken, and the average of all the samples, representative of the deposit. A representative sample must include all the material that lies between the lowest part of the area to be sampled, and the highest part of the area. This included sand, small pebbles, and larger rocks. The sample should include an equal amount of material from the top to the bottom of the area. Rocks which are too large to put into the sample, but lie in the sample zone should be estimated into the weight of the sample. In other words, don't throw anything out of the sample. The first sampling will be representative of the whole placer deposit. After certain information is learned from this sampling program, more exact sampling of certain areas may yield more specific information.

Sample size is very important, especially when dealing with a placer deposit. Large samples will consistently give more accurate results. Small samples of only a pound or two will usually vary drastically in the amount of gold they contain because of the nature of a placer deposit. Two small samples, taken next to each other, may assay at one dollar per ton and one thousand dollars per ton, respectively, just because one happened to have a large piece of gold in it. Larger samples help average these inconsistencies, and usually yield more reliable results.

A good placer sample should weigh at least fifty pounds. Larger samples are desirable, but may be limited by the method by which the samples will be concentrated.

A good sampling program may give you a definite answer to the economic reality of working your placer. It may also indicate what method will be best used, what equipment will be most effective, which layers of the deposit may be richer and which may have no value, where the hot spots of the placer are located, and where the poor areas are located. However, a definite decision often cannot be made after the first sampling. More sampling may be needed, or a larger scale sampling program may be set up right on the site. Several hundred yards of placer material put through a sluice box on the site represents a good sample and should answer many questions.

The results of the sampling program is not a final indication. These results must be compared with equipment costs, water availability, access to the deposit, pollution problems which may occur, royalty payments if any, and other economic factors. But a good sampling program does give you the best figure to put against these cost factors in determining if the deposit can be worked for a profit.

Before sampling can begin, you must decide how the samples you take will be processed to find how much gold they contain. The cheapest and easiest way is to take them to an assayer. But as stated before, small samples taken from a placer usually yield unreliable results.

An assayer uses a hot furnace to melt the samples he assays. The size of his furnace and equipment restricts the size of the sample he can assay. The assayer uses an Assay-ton System when weighing out his samples. An assay-ton is equal to 29,166 milligrams, or 29.166 grams, or about one ounce. Only this small amount of material is assayed from each sample.

This seemingly complicated system is necessary because of the different systems of weights and measures we use. Gold and other precious metals are weighed in troy weights and not in avoirdupois weights we commonly use. A troy ounce is about 1.1 times heavier than an avoirdupois ounce. Assay results are given in troy ounces per avoirdupois tons. There are 29,166 troy ounces in one avoirdupois ton. The assayers sensitive balances are scaled in metric weights, or grams and milligrams. He calls 29,166 milligrams one assay-ton. This weight unit is a convenient size for his equipment and convenient for him when he calculates his results. Each milligram in his assay sample then, represents one troy ounce. After he has completed his assay, he weighs the gold he has extracted from the sample. Each milligram of gold represents one troy ounce of gold. For example, if the gold bead he ends up with at the end of the assay weighs two milligrams, then the sample would assay at two ounces of gold per ton of material.

One ounce of material in a placer may vary greatly from another ounce. Even if a large sample is brought to the assayer, he can still assay only one assay-ton of it. It is impossible to mix a placer sample, so that the small amount of material finally used in the assay will be representative of the whole sample. A better way to process your sample is to concentrate the sample and then have the assayer find how much gold is in the concentrate. This is a far superior method because all of the gold will be recovered from the sample and sent to the assayer, who will assay all of the concentrate and produce a gold bead that represents all the gold in the sample.

Because gold is the heaviest material in the sample, it can be separated from the rest of the sample by its gravity. A small concentrate can be made which will contain all of the gold, the heavy black or magnetic iron, and a small amount of other heavy materials. This concentrate will be sent to the assayer who will extract all of the gold in the concentrate. This gold is then weighed and calculated against the total weight of the sample. For example, if the sample weighed one hundred pounds, and the gold that was recovered from it weighed 500 milligrams, then one ton of material would have twenty times as much gold in it, or 10,000 milligrams which is 10 grams. Since there are 31.108 grams in one troy ounce, the sample would assay at slightly less than a third of an ounce of gold per ton.

Before taking samples, find out what size is desired by the person who is going to concentrate them. Most concentrating equipment needs a certain minimum amount of material before it can operate properly. If the sample is too small, a reliable gravity separation may not be possible. On the other hand, samples should not be larger than necessary, mainly because the larger samples cost more to process. Very large samples may yield a large amount of concentrate. The assayer will not be able to handle large concentrates and they must therefore be reconcentrated. The more the sample is handled, the more it will cost to run the sample. Later in your sampling program large samples may be needed, but initially they should be kept close to the size recommended by the person who will concentrate them.

A small concentrating table (2'by4') works ideally for concentrating placer samples. Fifty to one hundred pounds of material put over a small table will yield a concentrate which the assayer can easily handle.

Before you start sampling, take a good look at the placer area. The nature of the river and its valley or canyon may tell you much about the placer deposit. Fast waters flowing through straight narrow canyons usually deposit very litte material. River bends and slower water usually indicate promising areas. Depressions and holes in the river course can have large amounts of gold in them. By studying the area of the placer, you may gain some information which will be useful in sampling the deposit. Some areas, however, are so changed from the time the material was deposited, that visual observation may be little help. Generally, the flatter the area of deposition, the more uniform is the distribution of gold through the material. The rivers which drain the east side of the Rockies in Colorado, for instance, deposit much material when they reach the plains. These deposits are fairly uniform and the further east one goes from the mountains, the less gold one finds, and the smaller the gold particles become.

Samples must be representative. This means the sample must include an equal amount of material from the top of the deposit to the bottom of the deposit. If the deposit is thin, holes may be dug by hand. Thicker deposits will require that you use some sort of equipment. A backhoe is usually used to dig sample holes if the material is not too wet. The holes should be dug to the bottom of the deposit if possible.

The only way to insure that a sample is representative is to take an equal amount of material from the top to the bottom of the hole. If the holes are deep, a bucket can be held against the wall, and sample material knocked into it with a prospect pick. Try to take the same amount of material as you move up the side of the hole to make your sample representative of the whole thickness. Pick a wall which is fairly vertical. Material which has sloughed in may have to be cleaned away from the bottom of the wall. Make the sample zone about a foot wide.

Be sure to include in the sample all the material that lies in the sample zone. Do not throw out large rocks because they are heavy or you know they do not contain any gold. They can be discarded later after the sample has been weighed. Making samples lighter than they really are makes them richer than they really are.

After you have completed taking the sample, label it and place a stake with the number of the sample on it at the hole.

When sampling in newly dug ground, especially in gravelly material, be very careful that the sides of the hole do not cave on you. If you have any doubts, brace the sides of the hole.

Sometimes a natural bank or a roadcut can be found on the site which exposes the deposit. Loose material should be cleaned off the bank and then the sample taken from the bottom to the top of the bank as mentioned before.

If the material to be sampled is very wet, it is difficult to take good samples. Holes dug in wet gravel quickly fill with water and cave in easily. The water in the hole makes it difficult to remove a sample which is representative, because the finer sand tends to be washed out by the water as the material is taken out of the hole.

Suction dredges can be used in wet material, but they are limited in the size of material they can pick up, and it is difficult to determine if the dredge is picking up a constant amount as it moves through the layers of the deposit. Again, the wet holes tend to cave, and the material picked up by the dredge may come primarily from the top layers of the deposit.

Sometimes an auger is used to drill a hole in the gravel. The material which is lifted by the drill is used as the sample. A pipe can sometimes be driven into the ground and then lifted and the material knocked out of it. However, the use of these devices is very limited because they can be used only in thin deposits which have no rocks in them.

One solution to sampling wet material is waiting until late summer or fall when the water level drops and then taking your sample.

Sometimes material which has been dug from the ground and piled up is sampled. This method is used in sampling material at sand and gravel plants to determine if enough gold is present to save as a biproduct. When sampling piled-up material, takes samples all the way around the pile and be sure that the samples are taken well above the base of the pile. A certain amount of concentration can occur as the material is piled up and again later as water from rain and snow-melt runs down the sides of the pile. The heavier material will tend to be found near the base of the pile. Samples taken right at the base of the pile may contain concentrated material.

Samples which contain a large amount of coarse material can

be screened, but only after the entire sample is first weighed. Samples are easier to carry and cheaper to ship if the larger rocks are taken out of them. But be sure the material taken out is clean and does not have finer material clinging to it.

An important part of sampling is to keep good records. Use a simple system to identify your different samples. Label all samples clearly and with waterproof ink. It is a good practice to put an identifying tag in with the sample as well as labeling the sample container. Always stake and label the exact spot where the sample was taken so that you can return to that place if you need to. Before you send your samples off to be run, be sure you have a written record of sample numbers and where each was taken.

Always put your samples in clean containers. Be especially careful that you do not get oil or grease in with a sample because if small particles are coated with these substances, they tend to float and this makes a gravity separation very difficult. If a small particle of gold is oil-coated, it may jump to the surface of the water and be carried into the tail rather than going into the concentrate.

The best containers in which to transport placer samples are metal or heavy cardboard drums with lids. A five gallon drum will hold seventy to eighty pounds of material. Large canvas sample bags or burlap sacks lined with two plastic trash bags — one inside the other — may be used.

As you sample, use a gold pan to gather additional information. With your gold pan you can determine the size range and shape of most of the gold particles. Visually stratified layers can be checked to see which carry value and which carry little or no value. This information will be important if the deposit is mined, but it may also be important to know before the samples are concentrated. For example, before the samples are concentrated they are classified, or screened to a certain size. If you have found particles of gold as large as 10 mesh, this information should be given to the person concentrating your samples so he will know that he must screen the sample no smaller than 10 mesh. If the particles of gold are very flat and flakey they will be harder to separate than more massive particles. Again, the person concentrating your samples should know this so he can run the sample accordingly.

Before samples can be concentrated they must be prepared.

This preparation includes weighing the sample, drying the sample, and screening the sample. The sample will be weighed before it is dried and then again after it is dried to find the percent moisture in it. Sometimes the percent moisture is found by weighing and drying only a small part of the sample and then using that percent for the whole sample.

Before a satisfactory gravity separation can be made, the material in the sample must be classified or sized. Rocks the size of apples and fine sand can not be concentrated together with any reliable results. Larger particles act like the heavier small particles and may even crowd them out of the concentrate. The best concentration occurs when the particles are all similar in size. The material in the sample will be screened to a size that will include all of the particles of gold if they are not larger than can be successfully concentrated. The oversized material from the screening will be carefully examined to see if any gold is present. If some is found it will be picked out and saved and later added to the concentrate. Screening can be done either wet or dry, usually depending on which type of material the concentrating equipment must have.

Laboratory equipment is constructed so that it can be easily cleaned. This is a very important factor when dealing with placer samples. If a few particles of gold are lost from a sample or if a few particles of gold are introduced to a sample the results may be very misleading. This equipment should be constructed so it is visually open and so it does not have cracks or corners or other places where material will gather. For example, wooden riffles, which are often used on concentrating tables, will swell and shrink as they become wet and then dry and will eventually crack. Small particles of gold may become trapped in these cracks as the wood swells and then later freed as the wood dries. The amount of gold in an average placer sample is so small that any gold lost or added to the sample will ruin the result of the test. The same care is also taken in preparing the sample. Containers and screens which are used are cleaned before and after use to ensure that none of the sample is lost and that nothing foreign is introduced.

The small laboratory concentrating table is probably the best suited device for concentrating placer samples. It can be easily cleaned if the deck and riffles are one piece or if the riffles are fastened to the deck in such a manner that there are no cracks or other places where material can be lost. The concentrating table can be tilted to compensate for light or heavy material and can easily handle a fifty to one hundred pound sample.

The products of a gravity separation on the concentrating table are the concentrate or "cons", the middlings or "mids", and the "tails". Sometimes the mids are included in the tails.

Simply, a gravity separation is to separate the heavy material from the light material. Rarely, however, is it as simple as it sounds when dealing with placer material. If all of the particles in the sample were the same size, shape and thickness, a very exact gravity separation would be possible. However, the different size and shapes of the gold particles can make them difficult to concentrate. Thin flakey particles behave like material with a lesser density and may show up with lighter material at the end of the separation. Massive particles of iron and other heavy material can even crowd these thin flakes out of the concentrate and into the mids or tail. These factors help determine how the pitch of the table is set to ensure a full recovery of the gold in the sample.

The sample is fed onto the high corner of the table and the separation occurs as the material moves toward the opposite end of the table. The back and forth movement of the table moves the material towards the opposite end and the pitch of the table moves the material at right angles toward the bottom of the table. It is somewhat like walking across a steep snow covered hillside. Each time you take a step forward you slip down the hill a small distance. By the time you get across the hillside you are lower than when you started. By the time the material in the sample reaches the end of the table it is lower. The lightest material has fallen most and falls off the end of the table near the bottom corner. The heaviest material has fallen least as it moved across the shaking table and it falls off the end of the table above the other material. The remaining material falls off the table between the heavy and light material.

Pans are placed at the end of the table to catch the material. The heaviest material is the cons, the lightest the tails, and the intermediate material the mids. Sometimes riffles are run the full length of the table near the top of the table to carry the concentrate. These "high grade riffles" keep the heavy material separated from the lighter material as it falls off of the end of the table.

This is a gas fired assay furnace. The samples to be assayed are mixed with certain reagents and melted in fire clay crucibles. One of the reagents is litharge or lead oxide, which is reduced to lead as the material in the crucible melts. This lead alloys with gold and silver present in the sample. After all the material in the crucible is melted, it is poured into a cast iron mold. After the material cools, the lead alloy is broken away from the rest of the material and put back into the furnace on a small piece of bone ash called a cupel. By heating the lead with the furnace door open the lead is oxidized back into lead oxide. After oxidation of the lead is complete a small bead of silver and gold is left on the cupel. This is the bead the assayer weighs to determine how much gold and silver are in the sample.

The material in the concentrate pan is removed and dried. Care is taken that all of the material is removed from the pan and that none of the concentrate is lost. The material in the mids and tails are carefully checked, usually with a gold pan, to see if there is any gold in them. The tails will comprise the largest bulk of the sample and should not contain any gold. The mids might contain some flaky gold, but should be small enough that they can be panned down and the gold in them added to the concentrate.

After being dried, the concentrate is carefully placed in a small bottle and sent to the assayer. The amount of material in the concentrate from a one hundred pound sample should not be too large for the assayer to fire in one cruicible. If the concentrate is large it can be carefully panned down to a more desirable size. The assayer will melt the material in the concentrate and produce a small bead which is usually composed of both gold and silver. Most placer gold has a fineness less than 950. This means that at least five percent of the weight of the gold is composed of impurities. Silver is the usual contaminate of placer gold.

The bead will be weighed and then boiled in a nitric acid solution. This is called parting and removes the silver from the bead. The gold bead which remains after parting is completed is weighed. This weight indicates how much pure gold is present in the sample. However, if the gold recovered from a placer is to be kept in its natural free form and not refined or melted, the weight of the gold-silver bead would be used to calculate the amount of placer gold per ton of material.

For example, if a placer sample weighing fifty pounds was concentrated and sent to the assayer and the weight of the gold-silver bead was 25 milligrams, and the weight of the gold bead after parting was 22.5 milligrams, then the material in the sample would assay at 0.03 ounces of gold per ton (22.5 milligrams 50lb times 40 — 900 milligrams 2000lb — 0.900 grams ton. 0.900 grams ton divided by 31.1 grams ounce — 0.03 ounces ton) Parting shows that ten percent of the placer gold in the sample is composed of silver, therefore the fineness of this placer gold is 900. If the assay was figured from the weight of the gold-silver bead the results would be ten percent higher.

After the sample results are studied a placer deposit will fall into one of the following categories: definitely not profitable, definitely profitable (very seldom), or a borderline case. The last case is common and the information gained during sampling and testing can help determine whether the borderline deposit can or cannot be mined for a profit.

A first step is to resample certain areas to confirm the test results. Always resample extremely hot or high grade areas, for these may be holding your sample average too high. If the results from different areas are fairly consistent, a few larger samples may be needed to give a more accurate answer. When sample results vary greatly from one area to another, a greater number of samples may be needed to give a more accurate average for the whole deposit.

The results of the concentrating tests should also be studied. Was all of the gold in the samples found in the concentrate or was some of it found in the mids or tails? This lighter gold, if it is present, may be difficult to save in an actual placer operation. Laboratory equipment is designed, and can be controlled, to make a very high recovery. Production equipment will not necessarily make such a high recovery, and gold which was difficult to save in the lab very well may be lost in the field.

The best way to get a final answer is to move temporary equipment onto the site and run several hundred tons of material. This is costly, but it should give definite answers. The equipment that would be needed includes a screening device, a sluice, and a front-end loader or suction device to pickup the material. If large rocks are present a grizzle will be needed ahead of the screen. The material is screened and then put across the sluice box and the concentrate saved. This type of sampling not only gives you large samples but also gives you an opportunity to attempt improving your recovery. Experimenting can be done by changing the pitch of the sluice, changing the amount of water running over the sluice, changing the screen size, trying different riffling in the sluice, and rerunning the tails under different conditions to see if any gold in them can be saved. These changes and others should give a good indication of just what you may expect to recover if the deposit is worked.

This is the lower end of the ragplant in operation. The ragplant empties into a screw classifier which is used to remove the slimes, or very fine material, from the sand.

The long structure extending from the right to the left is the ragplant. Sand and gravel are screened in the structure to the right. To the left is the screw classifier.

WORKING A PLACER

The size of the deposit will primarily determine how much money will be spent on equipment and what type of equipment will be used. Obviously, a deposit of several million tons will justify the expenditures of much more money than will an equally rich deposit of several thousand tons. Here we are concerned with the smaller operation where much of the equipment will be "used" equipment and some of the equipment will be built on the site.

The chief difference in working different placer deposits is the means by which the material is mined or moved to the concentrating equipment. If the deposit lays below or partially below the water table, a dredge or a dragline must be used to transport the material. The dragline is much more useful because it is not limited by the size of material it can pick up.

A hole is started with the dragline. The material from the hole is deposited near the concentrating equipment or piled up to be later transported to the equipment. The hole is enlarged and deepened until the bottom of the deposit is reached. When the dragline can reach no more material, it is moved back and the area where it was previously setting is mined. By constantly moving the dragline the whole area can be mined.

If the dragline feeds the screen directly, the screen and sluice must be made portable. If these pieces of equipment are built onto skids they can be moved by a caterpillar tractor so that they are always close to the dragline. A permanent sluice and screen may be set up and the material dug by the dragline transported to them. This method is not as economical because of the additional haulage of material to the concentrating area, and because the tails from the sluice must also be transported back to the pond being dug by the dragline. If the concentrating equipment is always close to the pond, returning the tails is not much of a problem. At least one hundred tons an hour should be moved with the dragline to make its use economical.

The suction dredge is limited by the size of material it can pick up. Unless the deposit is mainly fine material, the suction dredge will not be as effective as a dragline. The intake hose is moved over the surface of the bottom of the pond and sucks up material. A screen is placed over the end of the intake nozzle to keep rocks from plugging it up. If there are many large rocks present very little material will be picked up and the bottom layers of the deposit will not be reached. Suction devices work well only where the deposit is composed of sand and gravel with few large rocks.

A raft is needed to hold the sand pump which is the source of the suction. The material picked up may be pumped to the shore to be concentrated, or a screen and sluice box may be set up right on the raft. Unless a very large sluice is used, it is more economical to concentrate the material on the raft. The material will be transported only a short distance and the tails can be returned directly into the pond. The tails should be returned behind the raft or away from the suction intake so that they are not picked up again. If part of the deposit is above the water line the wall of the pond is continuously undermined and caved into the pond and picked up by the suction hose.

When working a deposit below the water line, always keep your pond a safe distance from the water course to ensure that the creek or river are never in danger of being muddied or flooded by your operation.

Dry deposits which lay above the water table can most easily be mined by using a caterpillar dozer to push the material to the concentrating equipment. Loaders can also be used, but they will not be able to move as much material. It will probably be advantageous to make the screen and sluice portable so that the distance the material is moved is kept to a minimum. This will also create more room to deposit tailings. If the equipment is stationary the tailings must be constantly moved away from the area.

The ultimate choice may depend on what equipment can be secured with the money that is available. No matter what equipment is used, however, a great deal of realistic planning must go into the use of that equipment to make its use as economical as possible.

The equipment needed at the concentrating area include a grizzle to remove large rocks from the placer material, a screen to screen the material to the size which will be put over the sluice, and the sluice box or rag plant. The most economical way to install this equipment is to have a gravity feed between each piece of equipment. The material which goes through the grizzle drops by its own gravity on to the screen intake and the screened material drops on to the sluice box by its own gravity. One disadvantage of having only gravity feed is that the unit may be too high to move easily. Conveyors are used to move the material where gravity feed is not used. These conveyors are also used to control the amount of material which is being fed to the screen and sluice. Conveyors and elevators may be needed, but their use should be carefully planned so they are not used needlessly.

If many large rocks are present in the placer material a grizzle will be needed. A grizzle is a row of heavy steel bars which screen out the very coarse rocks. Used mine rails turned upside down and welded to a heavy frame at each end makes a good grizzle. The finer material falls through the spaces between the bars and into a hopper below. The rocks which do not pass through the grizzle are shoved off to the side. The spacing of the bars will depend on the size of the screen which is used ahead of the sluice. This space is usually three to four inches.

The material in the hopper is fed directly into the screen or carried by a conveyor to the screen. The discharge rate of the hopper or the amount of material carried by the conveyor belt are controlled to give the correct amount of feed to the sluice box.

A trommel screen is used to screen the material to the size which will be put over the sluice. A trommel is a long rotating cylinder composed of one or more different sized screens. The intake end of the screen is higher than the discharge end so that material will move through the trommel. Material is fed into the intake end of the trommel and screened by the first screen. The size of this first screen is determined by the size of the material to be put over the sluice. Material screened by the fine screen goes directly onto the sluice box. If very coarse gold is present in the placer, the trommel will contain a second screen which will size material to be hand picked for large nuggets. The coarsest material passes through the trommel and is dumped out of the discharge end. The size of the screen or screens used will be determined during the sampling program.

The sluice box is the most important piece of equipment, because only here can the gold be saved. The sluice should be constructed of at least one quarter inch thick steel so that the bottom is flat and smooth. Adequate bottom bracing should be used to ensure that there is no buckling or rippling of the bottom. The sluice should be suspended so that the pitch of the box can be changed. Water coming onto the sluice should be controlled so that more or less water can be used at any time. The riffling and tie downs which hold the carpet in the bottom of the box should be constructed so a minimum amount of time is needed to clean the box. The width of the sluice will depend on how many tons per hour will be fed into the sluice. However, even for a very small placer, the box should be at least two feet wide. Wooden spacers can be bolted to the sides of the sluice box to make it narrower if necessary. The side of the box should be about ten inches high.

Successful gravity separation of gold on the sluice occurs only when the feed, pitch of the sluice, width of the sluice, and the amount of water coming onto the sluice are at optimum conditions. The only way to find these optimum conditions is to experiment. Generally, the rate of pitch of the sluice should be at least one and one half inches drop to the foot. If too steep, gold will be washed off the sluice, if too flat, sanding will occur. If the sluice is too wide for the volume of water available, sanding will occur and the material will move down the sluice in channels cut in the sand. If the sluice is too narrow the gold may not have a chance to drop to the bottom and be saved. Too little water will cause sanding. Too much water will prevent the gold from settling to the bottom of the sluice and being captured. The amount of material fed on to the sluice must be constant. Varying feed will cause the bottom of the sluice to alternately sand up and then be washed clean, and a poor gravity separation will occur.

A good gravity separation on the sluice requires that a constant amount of material is being carried down the sluice by a certain amount of water, and the pitch of the sluice is such that the heavier gold particles have a chance to drop to bottom and be captured; but the lighter material will not, and will be carried off the end of the sluice. By changing these variables these optimum conditions can be found.

Riffles are used on the flat bottom of the sluice to cause turbulences in the water, which drop the gold particles, and trap these particles once they are dropped. As mentioned earlier, steel punchplate riffling with indoor-outdoor carpet under it works well on fine gold (10 mesh and less). If coarser gold is present, additional riffling may have to be added to ensure that it is saved. Again, this is a case of trial and error, so make your riffling so it is easy to install and remove. If the sluice will save the coarse gold or the fine gold but not both, it may be necessary to use two sluice boxes. The first will have more severe riffling and will be adjusted to save the coarse gold. The tails from the first sluice will then be rescreened to a smaller size and put over the second sluice. This sluice will have a flatter pitch and be adjusted to save the fine gold.

The concentrate made on the sluice box is not a clean concentrate. It should contain all of the gold, but also black iron and small amounts of sand. If the placer operation is large enough to justify it, a clean up mill, as described earlier, may be built. Smaller operations may want to send their concentrates out to a mill to be cleaned. The mill will make pure placer gold, or sponge buttons, or gold bullion from the concentrates. The pure placer gold can be sold to anyone, the bullion and sponge buttons can be sold to the smelter. If you want to process your gold into bullion or sponge buttons yourself, a concentrating table, amalgamator, retort, and melting furnace will be needed. The concentrates from the table are put into the amalgamator and amalgamated, retorted and melted, as described earlier. If bullion is to be made a gold melting license is required. No license is required to make sponge buttons as long as one does not have more than 200 troy ounces of gold in sponge button form at any one time.

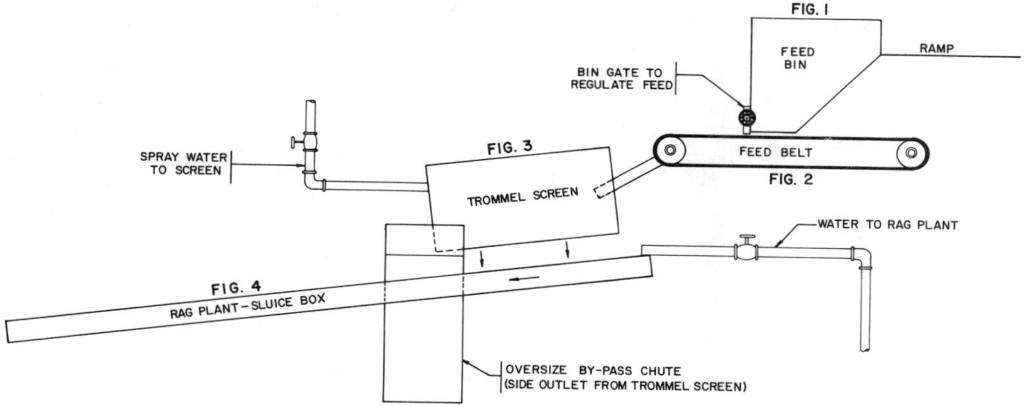

TYPICAL FLOW SHEET — PLACER OPERATION

(NO SCALE)

FLOW SHEET (when black iron also carries gold value)

In some placer operations with fine gold, where the black iron also carries free gold values, a different flow sheet has to be set up. Jigs are used to save this iron. The "hutch" product from the jig is put over concentration tables to clean the iron.

In a case like this, a rag plant ahead of the jigs and concentrating tables will scalp off 80 to 90% of the free gold recovered, with a very small amount of iron and gangue saved on the carpet.

This product from the carpet is cleaned on a small concentrating table. The table products are dried and put over the magnetic separator.

The jig-table concentrate is ground and generally cyanided if the tonnage is great enough, or is amalgamated if it is a small amount.

Another rag plant is used to protect the tails coming from both the tables and the jigs. This rag plant would not have to be cleaned as often as the rag plant ahead of the jigs, since most of the gold value has been scalped off. The product from this rag plant is processed the same as the rag plant at the head of the operation.

On page 29 the term "insul" is used. This is a long-used milling term, well understood by mining and milling men, but perhaps needing additional explanation. It probably should be "insol", for insoluable. Gangue, or impurity, is another synonym.

OPERATION INSTRUCTIONS

One of the most important things to put in effect in a placer operation like the one described, is to never allow the water to go over the rag plant when the feed has been shut off. The water, by itself, with no placer material going over the ragplant, will wash much of the free gold off and it will be lost.

When starting up the operation, the same thing must be in effect. Start the feed and water at the same time so that the gold on the carpets will be protected by the placer material in the water.

The carpets must be clean and the nap on the carpets not covered, so that the free gold can be caught in the nap and the finer gold go into the carpets.

In other words, when we say the carpets must be "clean", we mean you can not let the sand and iron build up on the carpet (like a sand bar).

With the right amount of feed, the right amount of water, and the correct pitch, or slope, on the rag plant, all of the sand and iron will be moving down the rag plant and the gold will be caught in the carpet (with a small amount of iron) when the carpets are cleaned.

All of the above must be done in the field, on a trial and error basis, until all three conditions are right.

A starting point could be 1¼" slope to each foot on the rag plant; then adjust up or down until the best condition is found. Water and feed are set the same way until best results are obtained. Once these results are obtained, a good, smooth operation can be maintained.

Fig. 1 (Feed Bin) should be large enough to assure a constant and even feed to the screen with enough cushion so that at least a half bin is maintained between truck loads (or however placer material is handled) from pit to feed bin.

Fig. 2 (feeder) There are several types of feeders that can be used the most common being a conveyor belt type with a gear-head motor to move belt very slowly so that the feed is fairly deep on the belt. An adjustable steel gate on the bin at feeder helps adjust to an even feed.

It is very important that an even and constant feed be maintained so that the right amount of water can be held at a steady rate. Also when the right amount of feed, plus the right amount of water is maintained, the exact pitch (or slope) of the rag plant can be set.

With all THREE of these conditions in control, the gold should settle in the carpet or the rag plant, and the sand and almost all of the black iron will go to tail and be eliminated.

Fig. 3 (Screen) A trommel or a vibrating screen can be used for classifying the material to go over the rag plant. Where there is quite a bit of large pebbles, a trommel screen is better, for it is more rugged and will stand up better. With finer material, a vibrating screen can be used.

On most placers with fine gold, a ¼" or ⅜" mesh screen is generally used. The fines will go to the rag plant and the oversize will come over the discharge end of the screen into a by-pass chute to waste.

The oversize material must be handled, so the screen should be elevated so that a stock-pile can accumulate and be moved with a loader, dozer or a conveyor belt.

Fig. 4 (Rag plant) This is a sluice box, 4 to 6 ft. wide and 35 to 40 ft. long. It has indoor-outdoor carpet in the bottom of the sluice with riffles on top of the carpet to catch and save the fine gold.

Several types of riffles can be used; punched plate, 3'X3'X3/16" with 1¼"X1¼" square openings punched out, works very well. Riffles made of 3/8" rebar welded into 3' by 3' sections, with 1"-1½" spacings can be used. There are also several other types of riffles which can be used.

It is very important that the riffle plates fit tight against the carpet. This is done by using ½" machine bolts sticking up about 3½" and coming through the bottom side of the sluice box. Bolts are welded into place with head of bolt on the under side sluice box. Nuts are screwed on to bolt, and tie-down tightened to keep riffles from moving and held tight against carpet. Wedges can be used between nut and tie-down to hold tie-down against riffles.

Tie-down plates are flat steel plates ¼"X3"X12' long. They are slotted to slip over bolt on top of riffle plate, and are tightened by screwing down nut on bolt, or by driving wedgw tight between nut and tie-down plate. (Shown in sketches on bolt and wedge set up.)

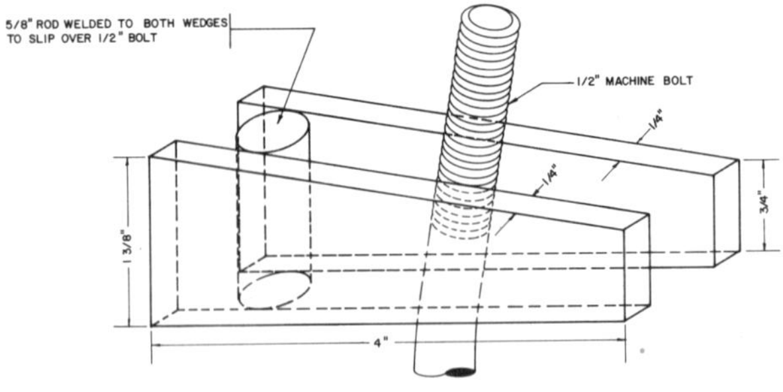

WEDGE

(SCALE · FULL SIZE)

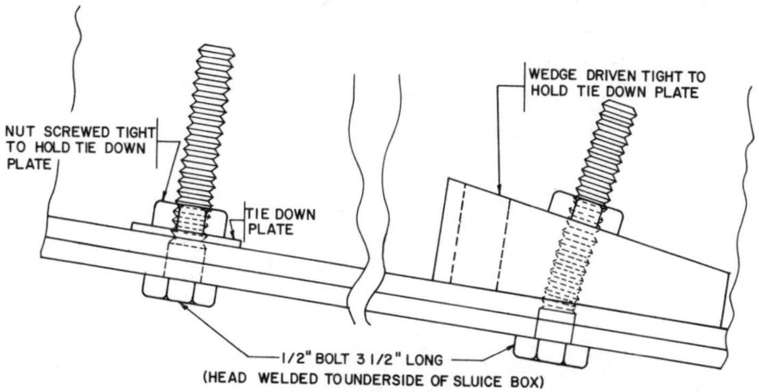

(NO SCALE)

BOLT AND WEDGE SET UP FOR RIFFLE TIE DOWN

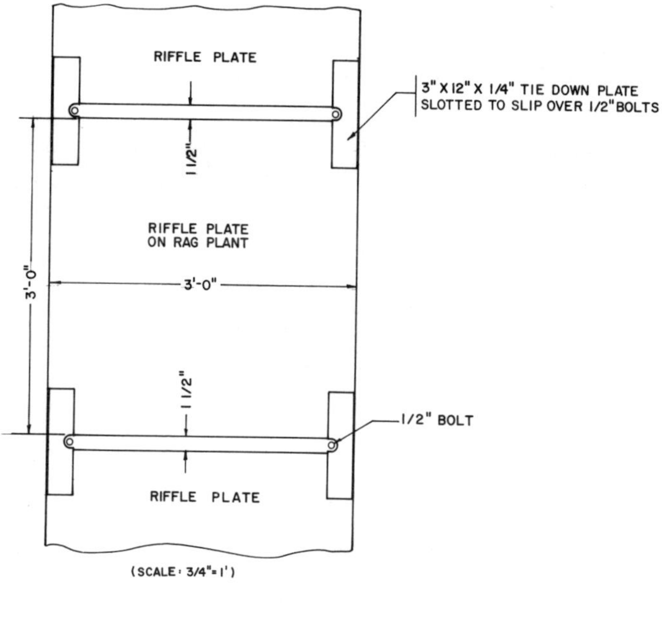

(SCALE: 3/4"=1')

AU — A GOLDEN FUTURE

Why point out these facts, ask a good many people, if the extraction of gold is thus possible? Why not just keep it all a deep, dark secret, do as well as you possibly can yourself, or maybe let a few close friends in on the deal? Why spill the beans? What will happen to the price of gold, for crying out loud, if too much is dumped on the market?

For one thing, it would be most difficult to place too much gold on today's market. Gold has become a great deal more than merely something of value. It has become a space-age, electronics-era raw material. For the same reasons, it is increasing demands by dentistry, jewelers, and coinage (some countries, you know, still mint coins made of the lovely stuff) it is also in demand by modern technology. Gold does not easily combine with other elements, and therefore is not readily corroded. It can last a long time under adverse conditions that would make parts made of other materials fall to pieces. It sounds somewhat strange that plumbing fixtures under certain conditions can more cheaply be made with gold, but it has become true. An ever-increasing demand for the durable, maleable, versatile metal is putting anyone who can produce the product in a most favorable position. It is a seller's market.

Such a market is in little danger of becoming flooded.

Too many people have doubts and bewilderment as to the weights used in weighing precious metals. Here are the straight facts, and beware of those who tell you differently. When weighing precious metals, such as gold, platinum, and silver, Troy weights are used exclusively. The United States Standard troy pound was copied in 1827 from the Imperial troy pound of England for the use of the United States mint, where it was deposited.

A chart of equivalent weights may be of use:

Troy Weight

24 grains — 1pwt (pennyweight)
480 grains — 20pwt or one ounce
5760 grains — 240pwt or 12 ounces (one pound troy)

The grain alone remains the same for avoirdupois weight. The ounce (oz.) and the pound (lb.) differ.

Avoirdupois Weight

1 ounce — 437.5 grains troy
1 lb. or 16 oz. — 1.2153 lbs troy

Metric Weight

31.108 grams — 1 Troy ounce
1000 grams — 1 kilogram — 32.146 Troy ounces

To change kilograms into troy ounces multiply the number of kg by 32.146.

To change grams into Troy ounces, multiply the number of grams by .03215.

SOME ANSWERS TO QUESTIONS SURE TO BE ASKED

This book is not presenting an easy get-rich scheme. It is a factual presentation of what has been done over many years of hard work by a lifetime of mining and milling to develop practical, profitable methods of gold extraction from formerly marginal or sub-marginal deposits. This is not theoretical. The facts presented here show what has been accomplished, a careful presentation of what can be done because it is being done.

Hopefully, this book will clear up many questions about the feasibility of getting into the **business** of extracting gold. It was written with that idea in mind.

Several questions should be answered before they are asked. First of all, Al McGowen is "retired". He is not going to go anywhere to check out people's potential placers. He is not manufacturing, nor does he currently have any ideas of manufacturing, any of the equipment he has developed over the years. It would consume too much of his time and be far too expensive to hand build, as he has with all his present equipment, the tools of his trade for others' use. If there is enough demand, he very likely would get some good tooling company to produce the equipment on a royalty basis. This would make it available at a reasonable price.

Jim McGowen, Al's son, is gradually assuming more and more of the work load. This father and son team, having more than enough work to do in their unique, ongoing business, cannot drop it to help others. They will try to answer questions, but only if they are reasonable. Al no longer does any sampling analysis work, as he often has done in the past, and Jim does not intend to do so, either. Nor can they get into the task of separating other people's concentrates, although they are interested in contacting additional sand and gravel operations.

These somewhat blunt statements are given here to keep Al and Jim from being swamped in a deluge of hopeful, well-meaning, interested inquiries. Al wrote this book as a possible help for those interested in better methods of gold extraction, not as an instrument of disruption to his retirement years. Please keep this in mind should you contact Al or Jim concerning the ideas presented within the pages of this book.

H. Glenn Carson